I0465686

The author is a civil engineer with 40 years' experience in the field of anti-corrosive treatment for concrete surfaces. He has to his credit more than 15 technical papers published and presented at national and international arena. It is his desire to share his applied knowledge with budding engineers who should also learn to protect concrete constructed under their eyes.

I would like to dedicate the book to my inspirational legendary guide—my beloved father whose iconic contribution to academics has brightened many careers, starting with me. I also dedicate the book to the budding engineers for enhancing their knowledge for protecting concrete produced under their eyes to overcome their ignorance for the subject which can cause chaos in any chemical industry.

Chirag K Baxi

# ANTI-CORROSIVE TREATMENT FOR CONCRETE SURFACES

AUSTIN MACAULEY PUBLISHERS™

LONDON • CAMBRIDGE • NEW YORK • SHARJAH

**Copyright © Chirag K Baxi (2020)**

The right of Chirag K Baxi to be identified as author of this work has been asserted by the author in accordance with Federal Law No. (7) of UAE, Year 2002, Concerning Copyrights and Neighboring Rights.

All rights reserved. No part of this publication may be reproduced, stored in a retrieval system, or transmitted in any form or by any means; electronic, mechanical, photocopying, recording, or otherwise, without the prior permission of the publishers.

Any person who commits any unauthorized act in relation to this publication may be liable to legal prosecution and civil claims for damages.

Austin Macauley is committed to publishing works of quality and integrity. In this spirit, we are proud to offer this book to our readers; however, the story, the experiences, and the words are the author's alone.

The age category suitable for the books' contents has been classified and defined in accordance to the Age Classification System issued by the National Media Council.

ISBN – 9789948250852 – (Paperback)
ISBN – 9789948250869 – (E-Book)

Application Number: MC-10-01-9050467
Age Classification: E

First Published (2020)
AUSTIN MACAULEY PUBLISHERS FZE
Sharjah Publishing City
P.O Box [519201]
Sharjah, UAE
www.austinmacauley.ae
+971 655 95 202

I would like to acknowledge the all-round, encouraging moral as well as commercial support given by Steuler Industrial Solutions (India) Private Limited because of which it has become possible to bring out this book.

I would also like to acknowledge the guidance of my senior colleague and path tracer, Mr. Rajesh Bhargava whose contribution in strengthening my basics on the subject and in building my moral spirit to bring out this book is exemplary.

# Table of Contents

# Chapter 8

# Preface

Civil Engineering education relates to different construction activities.

Whilst all aspects of design and construction are addressed in details under the prescribed curriculum disbursed over different years of the course, the subject of "Protection of constructed elements against variety of corrosive conditions" is often found missing from the same curriculum.

This subject is of vital importance. All constructed structures have to face variety of conditions under which they may deteriorate over a period of time. This deterioration can result in corrosion. Hence concrete structures which are designed and constructed for their functional life are required to sustain corrosive attack of many kinds.

In absence of basic classroom knowledge of this subject during studies, the student coming out of graduation fails to analyze such corrosive environments. It also brings in lack of confidence to deal with any mishap if it happens due to concrete or corrosion and there are all chances of getting misguided.

This book is a humble effort in the direction of imparting applied knowledge on the subject of corrosion and anti-corrosive treatments generally prevailing in the field to protect the concrete and structures to the students at their appropriate level of curriculum.

The contents of this book are to be well understood and implemented in field by integrating relevant information with each other. More often than not the solution would be a tailor-made one suited only to that particular situation for which past references might also not be available. Hence understanding the information given in the following pages in their required depth becomes necessary to have a good grip over the subject. This is significant because corrosive conditions are proportionately increasing everyday with industrial growth of the world. They have not remained limited to some chemical industries. Atmospheric conditions, chemical and effluent stagnation, bird droppings etc. are few of many such conditions which can cause corrosion to concrete or steel members.

At the end, I express my sincere thanks to Austin Macauley for their encouraging support in publishing this book and giving me an opportunity to contribute my applied knowledge for sharing it with budding engineers for enriching their engineering skills in the direction to make them as complete as possible. I owe my very special thanks to my good friend Mr. Abdul Latif for his untiring efforts in bringing out this book in Sharjah.

I wish the best of luck to all the students who would enrich their knowledge by studying this book.

**1st December 2018**
**Er. Chirag K. Baxi**

# Chapter 1

# Conditions Causing Corrosion of Steel and Concrete

Corrosion is a process which is responsible for deterioration of steel and concrete elements due to reaction of humidity and chemicals with iron particles for steel elements and cementitious substances for concrete elements. This deteriorating phenomenon results in loss of metal and concrete which is uncalled-for by any standard.

The process of steel corrosion is a chemical reaction between metal surface and humidity in presence of oxygen. This reaction results in formation of oxide salts which are not only responsible for increasing volume on the metal mass but also cause loss of basic metal because oxide salts is a degradation product of corrosion for metals. In case of reinforcement steel which is embedded in concrete mass, the process of increase in volume of steel due to corrosion generates positive pressure on concrete surface in absence of space to accommodate increased volume of corroded reinforcement steel.

In case of concrete surfaces, it is a reaction between cementitious elements and chemicals having high reactivity with cementitious elements. Chemicals causing heavy corrosion often show high reactivity with hydration products of cement. It results in depolarization of hydration products of cement due to which binding actions of cement start getting reduced and cement stats becoming ineffective. Very highly corrosive chemicals have tendency of imparting such a chemical reaction which almost dissolves hydration product of cement and causes immediate collapse of the affected concrete element.

In short corrosion is a process which invites loss of metal and materials from the built up elements. It is one of the most dangerous enemies of structures irrespective of its material of construction. Loss of metal and materials also affect material balance of nature because creation of new structures often needs consumption of fresh natural resources. If structures continue to corrode at this rate, that day is not far when industrial and domestic structures will struggle for survival in addition to loss of natural resources.

Hence corrosion is one of the biggest enemies for metal and materials of today and needs to be totally prevented to save metal, materials, nature, and natural resources. Fight against corrosion is need of this hour and is required to be a part of Civil Engineering curriculum so that civil engineers don't restrict application of their knowledge up to constructing the structures but also possess knowledge about maintaining health of structures constructed across the world and protecting them against corrosive attacks caused by different sources.

Sources causing corrosion broadly include chemicals, humidity, accumulation of duct and chemicals for long time, salty or saline atmosphere etc. All of them have high reactivity toward composition of metals and cementitious elements which results in their attraction toward such corrosive conditions and getting corroded.

Effects of corrosion are severe. Metal gets deteriorated due to corrosion. Cementitious structures like beams, columns, floors, drains, bridges, dams, roads, airports, residential structures, commercial complexes etc. get weakened due to corrosion. Industrial structures are exposed to such corrosive conditions at higher frequency as compared to residential or commercial structures. Hence industrial structures require more focused attention toward protecting them against attack of different chemicals. Effect of corrosion to industrial structures causes leakage of corrosive chemicals to subsoil. Floors get corroded. Supporting members like beams and columns for equipment are weakened. These conditions often end up in financial loss to the unit. Corroded slabs and roofs bring in leakage through them and disturb all activities in a production unit.

## Conditions Causing Corrosion:

It is absolutely necessary to know and identify the conditions which can cause corrosion so that appropriate protective measures can be designed to win over these conditions. This exercise ensures that the anti-corrosion treatment would effectively protect concrete against identified corrosive conditions. Majority of conditions causing corrosion are mentioned below.

A. **Atmospheric humidity:** Concrete surfaces which are exposed to atmospheric humidity are easily susceptible to algae and fungal attack which can corrode the surface in the long run.

B. **Corrosive chemicals:** Many chemicals have high corrosive value. They have tendency to react with concrete surface which can either soften the surface or erode it or cause a reaction to reduce strength and durability of concrete.

C. **Stagnation of corrosive liquid or gases:** Such conditions are encountered in chemical industries. At times, effluent coming out of a processing unit might contain some nutrient value and throwing it away could result in financial loss. Hence such effluent is often retained on floor or drains till it is recycled back for its reprocessing. Similarly inadequacy of effective ventilation might result in stagnation of corrosive gases within a building or production house. Such conditions cause severe corrosion on steel and concrete surfaces.

D. **Stagnation of wastes:** Sometimes sewage wastes from human, animal, or bird, dead bodies of animals or birds etc. remain unmoved for a long time. They decay and such decays also result in corrosion of concrete.

# Chapter 2

# Detrimental Effects of Corrosion

Corrosion is like white ant. It has very high detrimental effects on steel as well as concrete. As explained in earlier chapter, corrosion is a chemical process. It is a product of reaction between water present in humidity or chemicals and metal. Usually oxides are formed at the end of such reaction. This reduces metal content from the metal element. As a result, first deterioration to happen is reduction in thickness of the element. This is direct loss of metal by conversion of metal to rust. This is irreversible reaction. Hence rust doesn't get converted back to metal under any condition. Next step toward detrimental effect is reduction in sectional area and then in volume of the metal. Loss of metal continues in this process of deterioration.

Metal corrosion is serious because of following main reasons.

1.  Loss of thickness, area and volume of metal member makes it weak and incapable for functioning as designed and as desired. Lost thickness or area of metal is to be reinstated to restore that member for carrying load and stresses for which the same is designed. If this scenario is examined with an Industrial perspective, reinstatement of lost thickness of any metal requires stoppage of the unit where damaged metal member is located. This process causes production loss which is never a healthy business practice.
2.  Reinstatement of metal member largely involves safety precautions because it encounters many safety hazards like working at height, welding / cutting of

member, handling unsafe and corroded member to remove from its installed location, and new member to replace the damaged member.

3. Loose rust which is generated from metal corrosion might fall in an industrial vessel which is designed to handle fresh and clean materials. This is very dangerous as it can spoil quality of end product of that production unit.

4. There can be many more undesired consequences due to metal corrosion.

Similar detrimental effects are observed on concrete. Unlike in metals, humidity is not directly responsible for concrete corrosion. Concrete corrosion is caused mainly by corrosive chemicals consisting of harsh acids, alkalis, salts and solvents. They have tendency of reacting with hydration products of cementitious surfaces and hydration products of cement hydration get affected by this chemical attack.

Hydration of cement is a slow process which gradually continues till five years. It generates some complex intermediate chemical. It doesn't have a stable state of existence. Hence corrosive chemicals easily react with these intermediate chemicals generated during hydration process of cement. This often results in depolarization of cement and thereby basic properties of cement which is binding the concrete mass gets affected. When cement starts losing its binding property, it makes concrete a heterogeneous mass wherein (sand and aggregates) concrete ingredients no more remain bonded (connected with) to each other and concrete starts losing all of its properties.

Concrete corrosion is equally serious because of following main reasons.

1. Concrete is having good compressive strength but doesn't have good tensile strength. Compressive strength depends on compaction/consolidation of member. Better compaction/consolidation always makes the member strong and durable.

Heterogeneous concrete member loses its strength and stability and can't take any load or stress.

2. As mentioned earlier about effects of metal corrosion, reinstating concrete member is extremely hard depending on the location of the member i.e. whether its foundation, support beam, column, beam, lintel, slab or staircase west slab etc. As it is known to us, construction of concrete element requires water tight shuttering, reinforcement steel, and controlled concrete. In case of heighted concrete elements all these activities require extra high skills and precautions to ensure their construction with quality and safety.

3. If damage occurs in concrete member placed at ground floor, which is for drains or floor, initially their surface skin would start eroding. Coarse aggregates will start appearing on the concrete surface. This indicates that binding medium amongst coarse aggregates, fine aggregates, and other construction additives is getting lost and the member starts proceeding toward heterogeneity. Surface porosity also starts appearing on the surface of the affected concrete member.

4. If damage occurs in concrete member placed at elevated floors, in addition to the damages mentioned in point number 3, reinforcement corrosion will also appear on the concrete surface. This is visible as brown color traces on the concrete surface. Brown color is of rust which is a corrosion product of reinforcement steel placed inside concrete member.

# Chapter 3

# Need for Controlling Corrosion

Detrimental effects of corrosion on steel surfaces and concrete surfaces are examined in previous chapter. Corrosion is one of the most dangerous enemies for steel and concrete. It is required to be controlled to ensure long and desired life of the steel or concrete member and construction itself. Controlling corrosion necessitates knowledge about sources and conditions which can cause corrosion.

Sources and conditions causing corrosion for steel and concrete members were discussed in first chapter. They are summarized below.

1.  **Atmospheric humidity:** Steel surfaces which are exposed to atmospheric humidity are easily susceptible to corrosion though the humidity may not contain any harmful chemicals to additionally damage the steel surface.
2.  **Corrosive chemicals:** There are many chemicals which have high corrosive value. They have tendency of reacting with surface of steel and concrete and either soften the surface or erode the surface or cause such a reaction which might reduce strength and thickness of the member.
3.  **Stagnation of corrosive liquid or gases:** Such conditions are often met with in chemical industries. Liquid coming out of a processing unit which is called as effluent; might be found containing high nutrient value which is useful for production so throwing it away would end up in financial loss. Hence processing industries retain such effluent on floor or drains till it is recycled back for its

reprocessing till the nutrient value is recovered in manufacturing unit. Similarly inadequacy of effective ventilation might result in stagnation of corrosive gases within a building or production house. Such conditions cause severe corrosion on steel and concrete surfaces.

4. **Stagnation of wastes:** Sometimes human waste of any nature, animal/bird waste, dead bodies of animals or birds etc. remain laid at their disposed location for a long time. They decay and such decays also result in corrosion of steel or concrete member.

It is necessary to know or identify or find out corrosion potential of conditions or chemicals causing corrosion of steel or concrete member. This helps in deciding the appropriate and optimized actions for corrosion control. Some of those actions might need only fine tuning in surrounding conditions with disciplined approach while some situations would require systematically and scientifically designed corrosion protection treatment to effectively control corrosion and protect the steel or concrete member from getting corroded and losing their functional life.

Atmospheric corrosion is comparatively easy to control and to protect the member exposed to conditions causing atmospheric corrosion. However, corrosion caused by chemicals in surrounding atmosphere or under submerged condition requires accurate study of ingredients of these chemicals which will react with the surface of steel or concrete member and cause corrosion or deterioration. It is like knowing strength of enemy to calculate and arrive at the strength requirement to fight and win over the enemy. Such detail study ensures appropriate and tailor-made protective measure for each corrosive chemical - causing corrosion.

As the treatment needs to be cost effective, it has to be tailor made for each corrosive condition or corrosive chemical. It is not always true that highly expensive treatments are required to protect steel or concrete surfaces from highly corrosive conditions or chemicals. Following properties are important for materials considered for corrosion protection treatment.

i.      Inertness of material indicates the non-reactive index of that material when exposed to corrosive chemicals. It is also called as low reactivity. Inertness of material proposed for corrosion protection treatment is hence a significant property for materials to be considered for corrosion protection treatment.

ii.     Thermal expansion / contraction is another important property for materials to be considered for corrosion protection scheme design. Corrosion protection treatment is applied on the steel or concrete member; which already possesses thermal expansion / contraction coefficient. The magnitude with which it expands or contracts is fixed with respective change in atmospheric temperature surrounding the member. The material which to be considered for corrosion protection scheme should also have its coefficient of thermal expansion / contraction close to that of the mother surface to be protected.

iii.    Thermal properties of material being considered for corrosion protection scheme have an important impact on life of the treatment. These materials establish their corrosion resistant properties under chemical reaction between their ingredients. These reactions are normally exothermic which liberate heat during the reaction. This heat is required to be ventilated to atmosphere instead of sandwiching it which might generate internal cracks within the protection treatment mass.

iv.     Specific gravity and density of material with which corrosion protection treatment is to be carried out especially on vertical surfaces needs to be such that it does not exert additional load on the vertical member being protected.

v.      Selection of material for anti-corrosive treatment needs to be tailor-made for each corrosive condition.

Few situations may demand resistance to either only acidic environment, or only alkaline environment, or both acid and alkali present simultaneously; so corrosion protection treatment for such areas has to resist both these conditions. Few areas may also face abrasion in addition to chemical corrosion hence the treatment needs to be versatile to resist corrosive as well as abrasive effects.

vi.  Economics is one of the most essential factors in designing the anti-corrosive treatment under given conditions. Efforts are to be made to work out more than one option for the treatment and based on economics and cost spread over service life of the treatment depending on functionality and criticality of the structure to be protected, the system with optimum advantage is to be selected. Initial cost of few systems might be high but if that system gives more life than economical systems with low initial cost, then the system with high initial cost is preferred because its cost over the functional life makes it economical.

Design and implementation of protective measures for concrete.

Each protective treatment requires focused approach for its design. Factors affecting design of protective treatment are broadly summarized below.

i.  Factors which are likely to cause corrosion need to be focused at the time of designing the protective treatment for any environment. Sometimes more than one factor is simultaneously present at a single point which can cause corrosion. The protective treatment must be simultaneously resistant to all of them. This is a real challenge because acid resistance property differs from alkali resistance and when they are simultaneously present; it requires a treatment which is resistant to most of them.

ii.     Sometimes in addition to acid / alkali, the atmospheric or environmental conditions also play their role in increasing the damage caused by corrosion due to acid / alkalies. These factors are abrasion, wear and tear, thermal shocks, stagnancy, humidity, corrosive dusting from surrounding areas, chemical splashing, etc.

In case of single chemical to be resisted, design of the protective treatment requires only the consideration of temperature and concentration with which the acid / alkali is encountered by the surface. Then either by referring the corrosion resistance chart for each protective chemical or with experience of using appropriately proven protective treatment, it is selected.

In case of presence of more than one element causing corrosion, the treatment needs to be basically resistant the chemical having highest share in that mixture of corrosive chemicals. It ensures a fair protection of the surface. Other corrosive elements are to be examined for their process causing corrosion to arrive at the most appropriate protective treatment. The treatment as previously decided (to resist chemical having highest share in causing corrosion) must be compatible with treatment for other chemicals. For example, if acids and alkalis are present simultaneously or in tandem, treatment for each one of them is different but when a comprehensive treatment is to be designed for both of them to be simultaneously resisted, it requires a very précised examination of chemical reactions and only after ensuring the safety and durability of the protective treatment, it is to be finalized.

Conditions surrounding the area to be treated play vital role in designing the protective treatment. Humid atmosphere would retard the setting and curing process. Similarly high atmospheric temperature around the area under treatment would accelerate the setting and curing period. Hence manufacturer's recommendations as displayed on data sheet of each material selected for the treatment are to be carefully considered under these conditions at the time of designing the treatment. Few basic facts like "Alkalis at high temperature are more corrosive," "Few acids are more corrosive at lower concentrations" should also be addressed at the time of designing the protective treatment.

# Chapter 4

# Anti-Corrosive Treatments (Part I)

## Protective Measures for Concrete

It is a fact that prevention is better than cure. Same applies also for concrete. Protection measure for concrete is a multi-dimensional exercise. It requires very special skill and knowledge about corrosive index of the corrosive conditions and chemicals which can corrode concrete. Protective measures for concrete are broadly summarized below.

A. Acid resistant brick / Tile / Stone lining laid in appropriate mortar for bedding and jointing.
B. Joint-less / monolithic lining with appropriate chemical resistant resin with or without glass / fiber cloth reinforcement.
C. Factory-made hard sheet lining.

This chapter deals with Acid resistant brick / Tile / Stone lining laid in appropriate mortar for bedding and jointing.

A. **Acid resistant brick / Tile / Stone lining laid in appropriate mortar for bedding and jointing:**
   This is the most conventional method of protective measures for concrete protection. Here acid / alkali resistant bricks or tiles or stones are laid in bedding material which is resistant to the corrosive conditions. Joints of these bricks or tiles or stones are filled with mortar of highly resistant nature to corrosive conditions.

Factors affecting selection of materials for the work are mentioned below.

- **Joints on the floor:** Some floors are required to have minimum joints to ensure that any spillage on the floor doesn't penetrate below the floor. Some of the floors also require monolithic surface to ensure smooth movement of rolling stock (man or materials) on the same. At times, requirement of adequate jointing material and aesthetic look demands the work to be done with tiles of bricks which will have more joints.

- **Thermal expansion / contraction properties of materials:** Thermal expansion / contraction properties of materials used for anti-corrosive treatment are desired to have similar coefficient of thermal expansion and contraction as that of mother surface to prevent differential movement of mother surface and treatment medium.

- **Chemical resistance properties of materials:** This is one of the most important parameters for deciding the ant-corrosive treatment scheme. The material for bricks / tiles / stones and bedding / jointing mortar must have adequate resistance to the chemicals to which the treated surface is going to be exposed to. Inferior or nil resistance of treatment mortar to the chemicals to which the floor is going to be exposed to; the treatment will fail in short time.

- **Climatic conditions surrounding the work spot:** Climatic conditions like humidity, heat, or other state affect end performance of the anti-corrosive treatment. Mortars used for bedding and jointing of bricks / tiles or stones would set at lower rate in humid and cold conditions and vice versa, they will set fast in hot conditions. Hence mortar preparation and its application have to be in accordance with temperature and humidity surrounding the work spot.

- **Requirement of inter-liner:** There are some areas where strong acids / alkalis are to be encountered. Simultaneously it is also requirement that the floor has to be maintenance free for a long period. Under such circumstances, second line of defense is necessary. Inter-liner is provided in such cases below the final layer of bricks / tiles / stone. Material of inter-liner has also to be equally resistant to the chemicals to which the floor is going be exposed to.

Acid resistant brick / Tile / Stone lining is a stubborn and versatile conventional anti-corrosive treatment.

Usually the bricks and tiles are manufactured with ceramic material / filler; which is having high strength and also high inertness to many chemicals having different pH. They are laid in appropriate mortar for bedding and jointing.

Stone being naturally available material, only some particular spices of stones are used for this purpose which possesses acid / alkali resistance properties.

Mortar holds the key to successful performance of this lining. Bedding of this lining usually rests on bituminous layer consisting of either mastic or coating. Bituminous layer is preferred because of its fair acid resistance and humidity resistance properties. It provides a good priming function for its subsequent layers of treatment. Mortars of different materials have to be examined for ensuring the selection of right mortar for the chemical conditions to which it has to resist. Acid/alkali resistance charts of each material are available which helps in this selection. Normally it is ensured that mortar selected for bedding has good compressive strength in addition to chemical resistance.

**Acid / Alkali Resistant Brick / Tile Lining Joined with Suitable Mortar**

Mortars for bedding and joint filling:

## I. Potassium Silicate

Potassium Silicate cement confirms to IS-4832 part – I. It is resistant against a wide variety of acid conditions and covers almost the entire range of temperatures encountered in chemical plant process conditions. It withstands temperatures up to 800°C. It sets hard and dense, easy to use and apply. It is recommended for chemical plants, dye manufacturers, metallurgical industries, explosive manufacturers, textiles mills, storage battery manufacturers, plastic industries, oil refineries, fertilizers, etc. where mineral acids (except HF) such as chromic, nitric, hydrochloric, phosphoric, sulphuric (alone or in combination ) are used or dealt with. It is not recommended for alkaline and aqueous media.

II.   Sodium Silicate
It is similar to Potassium Silicate with some little changes in physical and chemical properties. Its usages are also for similar conditions for which Potassium Silicate is used.

III.  **Phenolic**
Phenolic is an organic resin based cement which is resistant to all acids (both organic and inorganic except certain strongly oxidizing ones), alkalis and even to chemicals with dilute concentrations. It is not suitable for strongly oxidizing acids, salt and alkalis. It is resistant to most organic solvents and can withstand temperature up to 1300 centigrade. It leaves no delay in construction because of its quick setting by chemical action. It is extremely dense, hence salt can't penetrate or crystallize.

IV.   **Furane**
Furane is siliceous organic cement. It is perfectly suitable for protective lining inside digesters dealing with severe acids like phosphoric acid, hydrofluoric acid, and other acid and chemicals. It has distinct advantage at room temperature for excellent resistance to erosion and sets into tough and hard bedding / jointing medium for brick / tile lining to resist attack of acids / alkalis. It is affected with change in temperature and humidity in its setting mechanism. Its setting slows down with drop in temperature which can be taken care by external warming up of site. It is used for acid resistant lining including carbon brick lining in chemical process industries, steel mills, tubing industry, electrolyzing plants, oil refineries, rayon plants, fertilizer industry, petrochemical plants, food, dairy, and dairy product industry, beverages manufacturers. meat and fish packing industry, breweries, jams and preservatives manufacturers, frozen food industry, candy/chocolate manufacturers, sugar refineries, vinegar and pickle manufacturers etc. It can also be used to join

stoneware terracotta sewage pipe lines carrying acid alkali waste and for acid disposal in neutralizing plants. Its thermal resistance increases with 2% addition of carbon by weight.

V. **Cashew Nut Shell Liquid (CNSL)**

CNSL is Cashew Nut Shell Liquid resin based cement having excellent chemical resistance properties. It is self-hardening, non-porous, and plastic in nature. It has distinct advantage of toughness with elasticity which makes it best suited for bedding and pointing for acid/alkali resistant brick/tile lining to mild corrosive conditions. It has unique advantage of its dilettance characteristics which enables it to penetrate inside cracks and crevices in bedding and joints of tiles/brick flooring. Its· Elastomeric nature makes it shock resistant. It provides toughness—not hardness and elasticity—not rigidity. Its highly viscous nature ensures total filling of joints between bricks and tiles. Other properties like toughness and elastomeric nature provide long life to the lining in totality. It is not adversely affected by oxidizing agents although it is not recommended for aromatic solvents and chloro compounds. It protects the surface against wide range of corrosive chemicals including alkaline solutions, mix of acidic and alkaline conditions, up to 90 C. It is also resistant to acids of low concentrations.

VI. Sulphur

This material is used for areas where weak acids and humid conditions prevail. Major limitation for using this material is its flash point. Hence areas where temperature is high or going to remain high with presence of inflammable materials around.

VII. **Epoxy**

Epoxy is a low viscosity solvent free mix of resin and hardener for protection of structure against heavy corrosive attack. These coatings are usually recommended in fertilizers, chemicals, and heavy chemicals, textiles, refineries, petrochemicals, foods, ships, pipes, and paper industries including water treatment plants, effluent treatment plants, and sewage drains. Its composition; being solvent free, it forms a non-porous and highly chemical resistant film as compared to conventional solvent based paints which have inherent porosity. It is free of fire and pollution hazards. It offers a brush applied coating covering with thickness of 250 to 300 microns in each coat and 3 to 5 mm thickness in mortar for screed, bedding, and pointing for tile/brick lining on floors and drains. It is compatible to be applied on steel, wooden, concrete, and even synthetic surfaces. The surface to be coated must be free of loose dirt, duct, oil, and grease.

VIII. **Isopthelic polyester**
IX. **Vinyl ester**

Vinyl ester resin is also a mix of resin and hardener generating highly resistant medium to wide range of acids, alkalis, oils, and fats. It is particularly recommended for its resistance against oxidizing agents and situations where rapid curing and hard setting is required. It adheres to all types of materials including keyed / glazed / unglazed bricks / tiles, metal, wood, and glass in bedding as well as pointing. It can also be used for protecting surfaces in contact with alkaline sub-strate. It has distinct advantage of rapid curing and attaining full chemical resistance properties within 48 hours of its application. These rapid setting and curing properties allow immediate use of floors, foundations, and drains after its application. It finds its application in places where strong and aggressive chemical corrosive conditions have to be encountered. It protects the surface against wide range of corrosive chemicals and oxidizing agents like Nitric Acid up to 30% concentration, chromic acid of all concentrations etc. It is also resistant to inorganic acids of low concentrations, hydrogen peroxide, Benzene, oil, and fats. It is completely resistant to many chemicals under controlled temperature and concentration.

X.  Carbon filled Furane

This is advanced version of Furane as explained above but with extra dose of carbon. This is very sparingly used for severe acidic conditions and where oxidizing agents play important part in chemical reactions in the plant where its use is being considered.

# Chapter 5

# Anti-Corrosive Treatments
# (Part II)

### Protective Measures for Concrete

It is a fact that prevention is better than cure. Same applies also for concrete. Protection measure for concrete is a multi-dimensional exercise. It requires very special skill and knowledge about corrosive index of the corrosive conditions and chemicals which can corrode concrete. Protective measures for concrete are broadly summarized below.

- A. Acid resistant brick / Tile / Stone lining laid in appropriate mortar for bedding and jointing.
- B. Joint-less / monolithic lining with appropriate chemical resistant resin with or without glass / fiber cloth reinforcement.
- C. Factory made hard sheet lining.

This chapter deals with monolithic lining and factory made hard sheet lining.

- B.1) **Joint-less / monolithic lining with appropriate chemical resistant resin without reinforcement**
  Monolithic lining has an inherent advantage of nearly zero joints which eliminates possibility of joint failure like that in case of brick / tile / stone lining.
  These linings are laid at site by preparing the material at site under strict quality control and taking adequate care for temperature compensation because mixing of

two or more ingredients of lining material are exothermic reactions.

Mixing of all components for this lining is also to be done in strict adherence with manufacturer's recommendations.

Thickness for this lining is a very important aspect. It is decided such that it not only resists chemical attack but also wear and tear to which the floor is exposed to.

At times, this is also pigmented as per the functional end requirement for the area where it is being applied. There are two actions to be complied by the lining before it is opened for use. The first is setting time which is the "touch dry" stage. Second stage is curing which is polymerization between all ingredients of the lining material and establishment of desired properties of the lining after which the lining can be opened for its use.

B.2) **Joint-less / monolithic lining with appropriate chemical resistant resin reinforced with glass / fiber or carbon fiber or polyester fiber**

This is similar to the previous treatment but as name suggests, the resin is reinforced with glass fiber or carbon fiber or polyester fiber. This reinforcement is extremely useful in improving mechanical strength as well as chemical resistance properties of the treatment. Glass fiber or carbon fiber or polyester fiber; which possesses high inert value against almost all corrosive chemicals is used for reinforcing the treatment which makes the treatment extra inert. Their fibrous texture reinforces the treatment which ultimately improves mechanical properties of the treatment layer which helps to resist abrasion and wear and tear on the floor. Few chemicals have tendency of reacting with resinous treatment and making it soft which can then be easily eroded from the surface. This phenomenon is well resisted and protected by providing fibrous reinforcement with the most suitable option.

## Properties of Different Fibers Are Discussed Below

### a) E glass fibers:

The use of glass fiber reinforced unsaturated polyesters is well known and is a major growth area. Many other polymers are, however, gaining ground in new and existing application areas where polymers replace traditional materials, such as glass fiber reinforced nylon and polyphenylene oxide. Composites have been reported in automobile radiator parts where they have fulfilled temperature and pressure requirements Ell. Nylon 6/6 resin, itself is a fairly tough material and is, in fact, among the toughest engineering plastics. It is resistant to corrosion and chemicals, but its application is limited due to low rigidity and strength, dimensional instability (i.e. higher coefficient of thermal expansion), and moisture absorption. On the other hand, E-glass fiber is extremely strong, having a low coefficient of thermal expansion and good rigidity but is brittle and susceptible to environmental attack. When they are combined they form a fiber composite with high strength, rigidity, toughness, and stability at elevated temperatures.

### b) Carbon fibers:

Carbon fiber possesses a unique combination of properties

- High strength-to-weight ratio
- Good rigidity
- Resistant to corrosion
- Conducts electricity
- Resistant to fatigue
- Good tensile strength but brittle

- Fire resistance/not flammable
- High thermal conductivity
- Low coefficient of thermal expansion and low abrasion
- Non-poisonous
- Biologically inert and permeable to X-rays
- Self-lubricating
- Excellent shielding against electromagnetic interference
- Relatively expensive
- Requires specialized experience and equipment for use
- High damping
- Electromagnetic properties

They are the stiffest and strongest reinforcing fibers for polymer composites, the most used after glass fibers. They have low density and a negative coefficient of longitudinal thermal expansion.

They have a high modulus of elasticity that results from the fact that the carbon layers tend to be parallel to the fiber axis. Fiber 'texture' is a term applied to this preferred orientation of the crystal structure. The modulus of elasticity of carbon fiber is higher parallel to the fiber axis than perpendicular to the axis. The stronger the "fiber texture," the greater the degree of alignment of the carbon layer parallel to the fiber axis. Carbon fiber with high fiber texture has high strength and high tensile energy absorption. The tensile energy absorption refers to energy stored in the fiber when the fiber is under tension with force and undergoes all extension or changes in length carbon fibers are very expensive and can give galvanic corrosion in contact with metals. They are generally used together with epoxy, where high strength and stiffness are required, i.e., race cars, automotive and space applications, sport equipment. Depending on the orientation of the fiber, the carbon fiber composite can be stronger in a certain direction or equally strong in all directions. A small piece can withstand an impact of many tons and still deform minimally. The properties of a carbon fiber part are close to that of steel and the weight is close to that of plastic. Thus the strength to weight ratio and stiffness to weight ratio of a carbon fiber part is much higher than either steel or plastic. Specific details depend on the matter of construction of the part and the application. In addition, the loading and boundary conditions for any components are unique to the structure within which they reside. The modulus of carbon fiber is typically 20 msi and its ultimate tensile strength is typically 500 ksi. High stiffness and strength carbon fiber materials are also available through specialized heat treatment processes with much higher values. In comparison with 2024-T3 aluminum, which has a modulus of only 10 msi and ultimate tensile strength of 65 ksi, and 4130 steel, which has a modulus of 30 msi and ultimate tensile strength of 125 ksi. Carbon fiber can be classified based on its properties; precursor fiber materials and final heat treatment temperature.

High heat treatment carbon fiber, where final heat treatment temperature should be 2,000° C and can be associated with high modulus type fiber. Intermediate heat treatment carbon fiber, where the final heat treatment temperature should be around or 1,500 °C and can be associated with high-strength-type fiber. Low heat treatment carbon fibers, where the final heat treatment temperatures is 1,000 °C. These are low-modulus and low-strength materials.

Most common uses for carbon fiber are in applications where high strength to weight and high stiffness to weight are desirable. These include aerospace, military structures, robotics, wind turbines, manufacturing fixtures, sports equipment, and many others. High toughness can be accomplished when combined with other materials. Certain applications also exploit carbon fiber electrical conductivity, as well as high thermal conductivity in the case of specialized carbon fiber. Finally, in addition to the basic mechanical properties, carbon fiber creates a unique and beautiful surface finish. Although carbon fiber has many important benefits over other materials, here are also tradeoffs one must weigh against. First, solid carbon fiber will not yield. Under load it bends but will not remain permanently deformed. Instead, once the ultimate strength of the material is exceeded, it will fail suddenly and catastrophically. In the design process it is critical that the engineer understand and account for this behavior, particularly in terms of design safety factors. Carbon fiber composites are also significantly more expensive than traditional materials. Working with carbon fiber requires a high skill level and many intricate processes to produce high quality building materials (for example, solid carbon sheets, sandwich laminates, tubes, and so on). Very high skill level and specialized tooling and machinery are required to create custom-fabricated, highly optimized parts and assemblies.

Joint-less lining also have its specialized uses other than anti-corrosive function.

Monolithic joint-less lining with epoxy or poly Urethane or other suitable resinous material is preferred for dust free environment, mechanically strong flooring, flooring with demarcation for specific purpose, and many more. Selection of resin for each application depends mainly on functionality and end requirement of the treatment.

c) **Factory made hard sheet lining:**

   As the name suggests, this lining is manufactured at manufacturer's workshop under quality control. These sheets are made up of complex polymers and co-polymers which have high resistance to respective aggressive corrosive chemical conditions. Accordingly sheet with suitable and appropriate polymer combination is selected for resisting the corrosive environment.

   Generally these sheets are available in 1.5 mm to 5 mm thickness.

   They are laid and fixed on the surface in the specified procedure as prescribed by the manufacturer.

   They are fixed to the mother surface with compatible resin or adhesive material.

Their fixing must ensure that no air bubbles are left between mother surface and the sheets because they form a weak plane from where failure can start.

Each sheet is joined with each other at their edges with welding technique as shown in the figure. These joints are tested for their perfect adherence by holiday testing technique.

**Joining Hard Lining sheets with PVC Welding Technique**

# Chapter 6

# Execution of Different Anti-Corrosive Treatments

Once the source of corrosion is established, corrosive conditions are identified, and material is selected for designing anti-corrosive treatment the most appropriate treatment for the conditions is to be finally decided. Factors affecting selecting an anti-corrosive treatment are listed below.

1. Surrounding conditions
2. Atmospheric conditions
3. Presence of corrosive gases or liquid at site
4. Site humidity
5. Time availability commensurate with curing period of the materials to be used for anti-corrosive treatment

Anti-corrosive treatments are broadly classified in following parts.

1. **Site laid lining:** This treatment is such that material to be applied under this treatment is to be prepared at site in the proportion as prescribed by the manufacturer and then it is to be applied at site. There are some important pre-requisites for this type of treatment. Except some very special conditions, the site is required to be dry and totally free from water or humidity before the treatment is laid. Site is also required to be free of all loose dirt, dust, oil grease etc. which can cause de-bonding of the treatment

from mother surface before taking up the treatment application. Only different components of the total lining are supplied form manufacturer's shop with instruction of their use are to be mixed strictly in prescribed proportion. Mixing the components in prescribed proportion is very important because ingredients polymerize after mixing with each other under a specific chemical reaction as per design of the final end product by the manufacturer. Slight mistake in proportion of mixing of the ingredients can end up in creation of some other material which was not desired and the entire scantily of treatment would go in vein. This type of lining is sub-divided in following sub-types.

### a)   Monolithic / seamless lining:

This lining, as name suggests, doesn't have any joints or seams. Once its laying starts, it is to be competed in one stroke instead of laying it in parts.

## Resin Based Lining:

Material used for this sub-type of lining is solvent less epoxy, solvent less poly urethane, etc. Normally solvent less resin is available in two components with individual ingredient present in them. Their mixing in the proportion as prescribed by the manufacturer is an exothermic reaction. Hence it generates heat from the reaction. This heat is to be appropriately handled such that it is not sandwiched between consequent layers of the treatment and crack the lining.

Severe conditions would require mortar of 3 to 5 mm thickness prepared by adding silica floor of quartz sand to the mixed resin. In case of mild conditions, only resin is applied in two or more coats over a coat of primer. Normal thickness of such lining is ranging from 400 microns to 800 microns.

Application of primer is essential for both the conditions because its ingredients help to establish bond between mother surface (plastered or concrete or steel surface) and subsequent coat of resin. This is very significant with respect to performance of the lining all throughout its functional life. All care of handling exothermic heat as explained earlier is to be meticulously taken while laying this lining.

## Fibre Reinforced Lining:

Material used for this sub-type of lining is glass / carbon / polyester fiber reinforcement for lining in suitable solvent less resin. This is used where abrasion resistance is required in addition to resistance to chemical corrosion. Fiber glass has very high inert value. It also provides required flexural and tensile strength to lining. It additionally improves impact resistance of the lining.

b) **Tile / brick / stone lining:**
This is the lining which consists of bricks, tiles, or stones of designed size and thickness which are fixed to the surface with help of bedding mortar. A designed space as specified in drawings and permitted by relevant Indian Standard will be left between each brick / tile / stone which will be filled up with suitable mortar.

Bricks, tiles, or stones used for this type of treatment are necessarily required to be having high resistance to corrosive conditions in which they have to function. Bricks and tiles are manufactured in factory using soil or clay having ceramic contents which are naturally resistant to acidic / alkaline conditions. Manufacturing process involves high temperature and pressure to establish their properties as per their norms set in respective applicable Indian Standard for their physical parameters like water absorption, acid / alkali resistance, tensile and flexural strength etc. Naturally available stones of some special category also possess chemical resistance properties. These are available in few mines in Rajasthan and Southern part of India. They are found to effectively resist the acidic or alkaline attack and they neither deteriorate nor allow any percolation of acidic or alkaline liquid to pass through them. They are largely preferred where it is required to have minimum joints in the anti-corrosive treatment because stones can be made available in larger sizes than bricks or tiles. However, few limitations of stones have also oflate been observed. These are natural stones and no artificial properties are added or imparted to them to improve these properties in them. Hence it is likely that some lot of these stones taken out from mines may have inferior resistance or no resistance to acidic or alkaline conditions. In some cases, chips of these stones in form of laminates are lost when they are exposed to acidic or alkaline conditions for a long time. Hence selection of basic medium of anti-corrosive treatment which is bricks, tiles, or stones is to be done after completely studying the site conditions to ensure the ultimate success of the treatment.

Mortar for bedding and jointing has a specific significance.

Basic function of bedding mortar is to provide bond between primer (which might consist of bitumen, bitumen mastic, or other suitable layer designed as per site conditions) and bricks, tiles, or stones. Bedding mortar is also required to have resistance to corrosive conditions like acidic / alkaline / humidity / simultaneous presence of both acidity and alkalinity etc. Bedding mortar also provides medium in which the bricks / tiles are laid in required slope, line, and level so it is necessary for this mortar to have adequate flexibility as well as setting properties to hold the layer of bricks, tiles, or stones as they are designed to remain throughout their functional life.

Mortar with which joints between bricks, tiles, or stones are filled is required to be effectively resisting the corrosive environment for resistance of which anti-corrosive treatment is designed.

Thermal properties of mortar are very significant with respect to setting and curing process of the anti-corrosive treatment. They should be within close proximity with thermal properties of bricks, tiles, and stones.

2. **Factory-made lining:** There are some corrosive conditions which demand "ready to fix" type of treatment. This is mainly required where site conditions are not conducive to providing conditions or pre-requisites for site laid lining. Being factory made material, it is manufactured under strict quality control so chances of quality of site laid lining getting adversely affected by either human error of mixing materials or laying the materials as per manufacturer's specifications are nullified in this case. Time required for preparatory works for site laid lining also gets reduced in this case as most of the work for manufacturing the ultimate material to resist corrosive conditions is already manufactured and it is just to be laid or fixed at site.

These are hard materials of designed thickness which ranges from 2 mm to 6 mm. Material of this lining consists of complex chemicals mostly in combination of more than one constituent.

Physical properties of this lining material like hardness, toughness, elastomeric properties etc. will be closely be required to be taken into account to ensure that thermal expansion and contraction effects or effects of corrosive shocks on the mother surface don't adversely affect the performance of lining.

Few commonly used materials for all the above mentioned categories of lining are listed below. These are few of the many available and used in market but those which are not added in the list are very sparingly used for ultra-special conditions in tailor-made format.

# Mortars for Bedding and Joint Filling:

I. Potassium Silicate
II. Sodium Silicate
III. Phenolic
IV. Furane
V. Cashew Nut Shell Liquid (CNSL)
VI. Sulphur
VII. Epoxy
VIII. Isopthelic polyester
IX. Vinyl ester
X. Carbon filled Furane

Execution of each job has two important steps to be followed. One is pre-treatment and another is post-treatment. Success of any protective treatment largely depends upon pre-treatment care and post-treatment care for the same.

1. Pre-treatment care primarily includes surface preparation before laying the treatment at site. Surface on which protective treatment is to be applied is required to be completely stable and strong. It must be free from cracks, loose dirt, and dust, oil and grease or any membrane which might de-bond or de-laminate the treatment from mother surface. It also should be free from humidity and moisture which retards setting and curing process of the treatment. Pre-treatment care also includes storing the ingredients to be use for the treatment at place as prescribed by manufacturer. At times it is necessary to open out the surface porosity to ensure that primer of the anti-corrosive treatment penetrates in initial layer of the surface so light grinding is also preferred for this activity.

2. Post treatment care is equally important for ensuring durability of the treatment. Subsequent to laying the treatment, it is to be ensured that two component resins used for jointing the bricks / tiles / stones resins

used for laying monolithic lining and adhesives used for fixing factory made hard sheet lining are set and cured only after which they are exposed to corrosive conditions. At times, site circumstances can't allow that time resulting in exposure of the treated area prior to curing of the resins. In such cases a sacrificial coat which can resist external attack of corrosive chemicals till the resins are cured is applied with an understanding that the sacrificial coat would get destroyed over a period of time but till then it ensures protection of resins till they get polymerized.

# Chapter 7

# Standard Specification for Acid Proof Lining

## 1. Scope

This section covers the specification for lining/coating to be provided over concrete and steel surfaces to protect them from the corrosive attack of chemicals in the form of leakages, spillages, overflows, washings etc. The scope of application is limited only to the external surfaces such as floors, pits, foundation sides and tops, encasing / coating to structural steel column, equipment supports such as skirts, legs, etc. and does not cover the lining to be provided inside the process equipment.

## 2. Applicable Codes and Standards

The following specifications, standards, and codes are applicable to this specification.

All standards, specifications and codes of practices referred to herein shall be the latest edition including all applicable official amendments and revisions. In case of discrepancy between this specification and those referred to herein, this specification governs.

IS 4832: Chemical Resistant Mortar (Parts I to III)
IS 4441: Chemical Resistant Mortar-Silicate Type
IS 4442: Chemical Resistant Mortar-Sulphur Type
IS 4443: Chemical Resistant Mortar-Resin Type
IS 4456: Testing of Mortar
IS 9510: Bitumen Mastic-AR Grade
IS 1580: Bitumen Compound

# 3. Types of Lining and Coating

Acid Alkali Resistant Brick-Lining
Acid Alkali Resistant Tile-Lining
Acid Alkali Resistant Epoxy-Lining
Acid Alkali Resistant Monolithic-Lining

# 4. Material Specifications

## 4.1. Bituminous Primer Coat

Bitumen shall be of Acid Resistant Grade conforming to IS: 3384. It should be suitable for either cold or hot applications over concrete and /or steel surfaces.

## 4.2. Bituminous Mastic Layer

The mastic for hot application over the primer coat shall be of Acid Resistant grade conforming to IS: 9510. The general purpose Acid Resistant Mastic shall be used for floors, drain, or trenches of depth not more than 80 to 100 mm. The mastic layer shall be used only as an inter-liner, before application of bricks/tiles.

## 4.3. Hard sheet Lining in Place of Mastic Layer

Normally, this is used only as inter liner before the application of brick and tile lining. Minimum recommended thickness of this lining is 6 mm. The material to be used shall be from the approved manufacturer. A coat of emulsified bituminous or other suitable bonding material shall be applied over this lining to make it compatible with subsequent layer to achieve the necessary proper bond between both.

## 4.4. Chemical Resistant Bricks/Tiles
### 4.4.1. Bricks

They shall be of Acid resistant quality, conforming to IS: 4860 and of approved make. The size shall be 230 x 115 mm. and thickness shall be 38, 50, 65, or 75 mm, as specified in the drawing and schedule of item description. Its surfaces shall be even and free from undulations, cracks, holes, pits, etc. Their dimensional tolerance shall be 2%.

#### 4.4.2. Tiles

Tiles shall be of ceramic material with vitrified surface conforming to IS: 4457 with not more than 20 mm thickness.

### 4.5. Cementing Materials For Bedding, Jointing, and pointing

The cementing materials to be used for bedding and jointing shall be of chemical resistant quality and conforming to the respective IS-Specifications. The material shall be one of the following types of cements and as specified in the drawing and schedule item description. The shelf life of the various materials shall be carefully observed as specified by the manufacturer.

#### 4.5.1. Potassium Silicate Cement

It is used as bedding as well as jointing material. Jointing on sides shall be having an average depth of 20 mm. Minimum thickness of bedding is 10 mm. Mixing of its resin and hardener and application shall be as per manufacturer's specifications.

#### 4.5.2. Phenolic Resin Based Cement

It is used as bedding and jointing material or pointing material.

#### 4.5.3. Furane Resin Based Cement

It is used as bedding and jointing or pointing material.

#### 4.5.4. Cashew-Nut Oil Resin Based Cement

It is be used for bedding and jointing or pointing material.

#### 4.5.5. Vinyl-Ester Resin Based Cement

It is be used as bedding and jointing or pointing materials.

### 4.6. Epoxy Based Cements

They are used as bedding and jointing or pointing or screeding / coating materials as specified in schedule items and construction drawings. All ingredients of Epoxy formulations to be used shall be of approved make. Minimum thickness to be used for different applications shall be as follows:

A. Primer Coat – Minimum DFT of 50 microns or as per the manufacturer's specifications, whichever is more.
B. Unreinforced application brush or trowel is carried out by adding Silica or carbon powder as filler materials. Min. thickness per coat shall be 150 microns.
C. Unreinforced monolithic screed – Thickness for the same shall be 3 to 5 mm.
D. Reinforced Brushable Application.
   Brushable Epoxy application, reinforced with appropriate fibrous material of either E glass or carbon fibers or polyester or polyethylene fibers; shall have a min. thickness of 200 microns for a single layer consisting of two flood coats of epoxy with suitable fabric sandwiched in between. Number of such layers shall be decided as per schedule of item specifications and construction drawings.
E. Reinforced Monolithic Epoxy Screed
   Min. thickness of one layer consisting of two sub-layers of monolithic epoxy mortar application with fiberglass tissue fabric reinforcement sandwiched between the two shall be 4 mm and number of such layers shall be decided as per schedule of item specifications and construction drawings.

## 4.7. Fillers and Reinforcement – Fabric
### 4.7.1. Silica or Carbon / Graphite Powder
The powder particle size shall be passing through BS-sieve 100 for Quartz silica powder and carbon / graphite powder and shall be clean and free from impurities and moisture. The proportion of filler material by weight to be added to resin shall be as per manufacturer's specification and shall be different for different applications.

### 4.7.2. Fibre Glass Tissue Fabric

This shall be of approved make. The grade to be used shall have min. unit weight of 450 gm / sqm. of the fabric

.

# 5. Methodology for Acid Resistant Brick-Lining

## 5.1. Surface Cleaning and Preparation

Concrete or steel surface to be protected shall be carefully examined to ensure evenness, freedom from cracks / holes / undulations, and shall be thoroughly cleaned of all loose particles, dust and preferably by using dry compressed air. In underground pits and trenches, the leak test shall be insisted upon and shall be examined for dampness, seepage etc. on the bottom or walls. Any defects in concrete or pits and dampness shall be rectified. The surface shall be made dry before application of primer coat. If the surface is not levelled to the required slopes and grade, the same shall be got done by using cement mortar screed, properly set and cured. One coat of Bitumen primer as specified shall be applied on the clean dry surface of the screed, after ensuring the quality acceptance of the surface to be protected and obtaining approval thereof. After the application of Bituminous Primer a layer of acid resistant mastic shall be laid over it. The mastic layer shall confirm to IS: 9510 and shall be of acid resistant grade. The thickness of this layer is decided as per the functional need. In HF-Conditions carbon filled material shall be used in place of Quartz powder.

## 5.2. Application

The bricks are set in full mortar bed and jointing is done as specified. The bricks shall be coated with the mortar on side forming bed and joints and then placed in position by properly pressing and squeezing out the excess mortar to ensure 100% application over brick surfaces. The brick laying shall be done in one of the ways specified as follows with respect to type of cement used in bedding and jointing.

a) **Potassium Silicate Cement**

This shall be used only as a bedding material and not for pointing work. The mortar bed shall consist of approved Potassium silicate cement and prepared conforming to IS: 4832 (Part-1) and IS: 4456 (Part-1). Bricks shall be set in position after application of mortar of appropriate consistency, over bottom and sides, of the brick and then properly pressed in to its place ensuring minimum thickness of 10 mm for bedding and 5mm for jointing having 20mm deep joint, open from top of pointing work. All the loose particle and impurities from the joints shall be removed and pointing shall be done using selected resin based cement. The joints shall be arranged in such a way that there is a continuous joint in the longitudinal direction of bricks and in the direction of slope. They should be staggered in the perpendicular direction of slope and in the cross direction of longer side of bricks.

b) **Phenolic Resin Based Cement**

This shall be used as bedding and jointing material or pointing material. The mortar conforming to IS: 4832 (Part-2) shall consist of approved phenolic resin based cement and filler of quartz silica powder. Thickness of mortar shall be 6mm and bricks shall be laid with 3mm wide joints filled with phenolic resin cement completely including flush pointing with the same cement. Arrangement of joints shall be as described under Alternative 1 above. The mortar shall be applied with trowel over bottom and sides of the brick before placing it in position.

c) **Furane Resin Based Cement**

All specifications shall be as per Alternative 2 except for the cement, which shall be approved Furane resin based in place of phenolic resin based.

**d) Cashew-Nut Oil Resin Based Cement**
All specifications shall be as per Alternative 2 except for the cement, which shall be approved cashew nut oil resin based in place of phenolic resin based.

**e) Vinyl-Ester Resin Based Cement**
All specifications shall be as per Alternative 2 except for the cement, which shall be approved vinyl ester-based resin.

# 6. Method of Application for Acid Resistant Tile Lining

Procedure for tile lining shall be exactly the same as for brick laying except that the thickness of the tile shall be generally limited to max. 20mm or as specified in drawings, whichever is lesser. Cement used for brick lining are applicable for tile laying except for potassium silicate cement which is not applicable in case of tile lining.

## 6.1. On Curbs and Wall Connections

The lining and mastic shall be taken and terminated horizontally to cover the top of curbs and walls. Further continuation of lining over outside surfaces of curbs and walls shall be as per construction drawing.

## 6.2. Lining over Equipment Foundations

When lining is done over equipment foundations the portion of bolt holes within the thickness of tile lining shall be filled with the same cement as is issued in the lining. Pockets, if left for bolts, shall be grouted up to unfinished top of concrete with cement grout prior to the lining work. This shall be in scope of others. Wherever a pipe nozzle is penetrating through wall lining shall be laid to ensure that there are no crevices, gaps around the nozzle by properly filling up the cement in the annular gap around the nozzle and puddle flanges.

# 7. Method of Application for Acid Resistant Epoxy Linings

The following specification covers the requirements for the application of epoxy-based lining.

## 7.1.

For materials that shall be limited to areas where maximum temperatures are below 90°C or as specified by the manufacturer, whichever is less. Procedure for epoxy lining shall be exactly the same as for brick laying. The material to be used must be from reputed and standard manufacturer. They shall be stored and used as per manufacturer's instructions because epoxy resin materials require great care in handling and application.

### 7.1.1. Brushable Application Without Fiberglass Reinforcement

**a) (a)Primer Coat**

One coat of primer shall be applied over the surface ready for lining, with a clean brush using a mixture of resin and hardener mixed and prepared in proportion as specified by respective manufacturer according to pot-life as per manufacturer's specification.

**b) (b) Finish Coats**

A mixture shall be prepared to the brushable consistency by mixing resin, hardener, and quartz silica sand in proportion specified by the manufacturer. Quantity of quartz powder in a batch shall be adjusted so as to get a proper brushable consistency. Min. 2 coats shall be applied with drying and curing time interval between the two coats as specified by the manufacturer.

### 7.1.2. Reinforced Brushable Application

Brushable application with fiberglass tissue fabric reinforcement all specs. shall be the same as per Alternative-1 except the fiberglass tissue. Reinforcement shall be laid after two coats of epoxy formulation. The fabric shall be laid and pressed in position as to remove all air pockets, waviness, folding, wrinkles, etc. in the fabric layer. Nos of such sandwich layers shall be as per construction drawings.

## 7.2. For Temperatures of 100 Deg C to 150 Deg C
## The Following Specifications Shall Be Used, All Materials Shall Be of Approved Make

### 7.2.1. Material
Material shall be similar to previous applications except for special one having temperature resistance.

### 7.2.2. Surface Preparation
Surface preparation shall be described in Para 5.1. Care shall be taken to see that the surface is free from moisture. The slab is tested for moisture content and a slab with less than 12% moisture is considered suitable for application of the epoxy lining. Testing shall be carried out by covering an area of 1 sq. mtr. With a polythene sheet with edges sealed against external moisture for a min. period of 7 days. If moisture collects on under surface of the sheet then the slab is not considered suitable for the application of epoxy coating. It should be ensured that moisture does not enter the underside of polythene sheet through sources other than evaporation. When the time available for testing is short then the following two methods are used:

a) The surface shall be covered with a nonporous rubber mat for twelve hrs. If moisture collects on the under surface of the mat then the coating shall not be applied.

b) A few granules of calcium are placed over a small area and covered with a dry glass plate and the edges are sealed to prevent the entry of moisture from outside. If the granules are dry after 2 to 3 hrs. then the slab is suitable to receive the coating. Alternatively, an approved moisture meter may be used for measurement of moisture percentage.

## 8. Mode of Measurement

Payment shall be on M2 basis of finished area. The rates quoted should be inclusive of materials, labor, supervision, transport, all applicable taxes and duties, wastage, guarantees, profits, and all other incidental expenses.

# Chapter 8

# Chemical Resistant Chart

| Sr. No | Substance | Epoxy | Polyester | Phenolic | Furane | Cashew Nut Shell Liquid (CNSL) | Sodium Silicate | Potassium Silicate | Sulphur |
|---|---|---|---|---|---|---|---|---|---|
| 1 | 2 | 3 | 4 | 5 | 6 | 7 | 8 | 9 | |
| | **Acids:** | | | | | | | | |
| 1) | Acetic Acid 10% | R | R | R | R | R | | | |
| ii) | Chromic Acid 10% | N | R | L | N | L | | | |
| iii) | Hydrochloric Acid (Conc) | R | R | R | R | R | R | R | R |

| | Col 1 | Col 2 | Col 3 | Col 4 | Col 5 | Col 6 | Col 7 | Col 8 |
|---|---|---|---|---|---|---|---|---|
| iv) Hydrochloric Acid 40% (See Note 2) | N | N | R | R | R | | | |
| v) Lactic Acid 2% | R | R | R | R | R | | | |
| vi) Nitric Acid 10% | L | N | L | N | L | | | |
| vii) Nitric Acid (Conc) | N | N | N | N | N | | | R (<40%) |
| viii) Phosphoric Acid 10% | R | R | R | R | R | | | |
| x) Sulphuric Acid 40% | R | R | R | R | R | L | R | |
| ix) Sulphuric Acid (Conc) | N | N | L | N | N | N | R | R(<70%) |
| **Alkalis:** | | | | | | | | |
| i) Amonia 0.880 | R | N | L | R | R | | | |

| | | 1 | 2 | 3 | 4 | 5 | 6 | 7 | 8 |
|---|---|---|---|---|---|---|---|---|---|
| ii) | Sodium Hydroxide 40% | N | N | N | L | R | L | N | R |
| iii) | Sodium Carbonate | | N | N | R | R | R | L | R |
| iv) | Calcium Hydroxide | | | | R | R | R | N | R |
| | **Salt Solutions:** | | | | | | | | |
| i) | Salt Solutions (acidic) | R | R | R | R | R | R | R | R |
| ii) | Salt Solutions (alkaline) | N | N | N | R | R | R | L | R |
| | **Solvents:** | | | | | | | | |
| i) | Aliphatic Hydrocarbons | R | R | R | N | R | R | R | R |
| ii) | Aromatic Hydrocarbons | R | R | R | N | R | R | N | L |

| | Col 1 | Col 2 | Col 3 | Col 4 | Col 5 | Col 6 | Col 7 | Col 8 |
|---|---|---|---|---|---|---|---|---|
| iii) Alcohols | R | R | R | R | R | R | R | R |
| iv) Ketones | L | N | L | R | R | R | R | R |
| v) Chlorinated Hydrocarbon | L | L | R | R | N | R | R | R |
| | | | | | | | | |
| Wet Gases (oxidizing) | N | N | N | N | N | | | |
| Wet Gases (reducing) | R | R | R | R | R | | | |
| Mineral Oils | R | R | R | R | L | | | |
| Vegetable Oils and Fats | R | R | R | R | L | R | R | R |

Note 1: R= Generally recommended
        L= Limited use (occasional spillage)
        N= Not recommended

Note 2: Carbon and graphite filters should be used for hydrofluoric acid service

www.ingramcontent.com/pod-product-compliance
Lightning Source LLC
Chambersburg PA
CBHW070127230526
45472CB00004B/1448